W9-BRV-059

DATE DUE

THE STORY OF
NUMBERS
AND
COUNTING

by Anita Ganeri

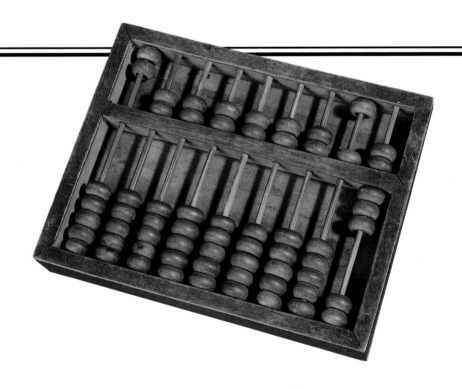

OXFORD UNIVERSITY PRESS

Published in the United States of America by
Oxford University Press, Inc.
198 Madison Avenue
New York, NY 10016

Oxford is a registered trademark of Oxford University Press

Published in the United Kingdom by Evans Brothers Limited
2A Portman Mansions
Chiltern Street
London W1M 1LE

Printed in Hong Kong by Wing King Tong Co. Ltd

Library of Congress Cataloging-in-Publication Data
Ganeri, Anita.
  The story of numbers and counting / by Anita Ganeri.
    p. cm. — (Signs of the times)
Includes index.
ISBN 0-19-521258-4
  1. Numeration — Juvenile literature. 2. Mathematics — Juvenile
literature. [1. Number systems. 2. Mathematics.] I. Title.
II. Series: Ganeri, Anita, 1961- Signs of the times
QA141.3.G35 1996
513.2 — dc20   96-14759
                                                   CIP
                                                   AC

Acknowledgments

Editor: Karen Ball

Design: Neil Sayer

Illustrations: Hardlines

Production: Jenny Mulvanny

With special thanks to David Bryden and Peter
Patilla who acted as consultants on this title.

Acknowledgments
The author and publishers would like to thank the following for permission to reproduce photographs:
Front cover (main picture) The Bodleain Library, (top left) AKG London, (top right) Paul Schemt, Eye Ubiquitous,
(bottom left) AKG London, (bottom right) The Image Bank
Back cover Mary Evans Picture Library
Title page The British Museum
page 6 (middle) Luiz Claudio Marigo, Bruce Coleman Limited, (bottom) The Science Museum/Science and Society
Picture Library page 7 (left) Jerome Yeats, Science Photo Library, (right) Mary Evans Picture Library page 8 (top) The
British Museum, (bottom) British Museum, e.t. archive page 9 Alain Compost, Bruce Coleman Limited page 10 (top)
Nicholas Devore, Tony Stone Images, (bottom) Liverpool Museum, Werner Forman Archive page 11 Paulo Koch, Robert
Harding Picture Library page 12 (top) Hugh Sitton, Tony Stone Images, (bottom) Nicola Sutton, Life File page 13 Tony
Stone Images page 14 (top) Mary Evans Picture Library, (bottom) Peter Newark's American Pictures page 15 Victoria
and Albert Museum, The Bridgeman Art Library page 16 (top) Ronald Sheridan, Ancient Art and Architecture
Collection, (bottom) J-L Charmet, Science Photo Library page 17 (top) Mary Evans Picture Library, (bottom) Topham
Picture Source page 18 Mary Evans Picture Library page 19 (top) Science Photo Library, (bottom) Mary Evans Picture
Library page 20 (top) British Library, The Bridgeman Art Library, (middle left) Adrian Davies, Bruce Coleman Limited,
(middle right) Ancient Art and Architecture Collection page 21 (top) The Hutchison Library, (bottom) Henry Tse, Tony
Stone Images page 22 (top) Ronald Sheridan Ancient Art and Architecture Collection, (bottom) Bibliothèque Nationale,
The Bridgeman Art Library page 23 (top) e.t. archive, (bottom) Mark Harwood, Tony Stone Images page 24 (top) The
Science Museum, Science and Society Picture Library, (bottom) Fiona Good, Trip page 25 (left) Mary Evans Picture
Library, (right) Honourable Society of the Inner Temple, e.t. archive page 26 (top) The Science Museum, Science and
Society Picture Library, (bottom) Topham Picture Source page 27 (top) The Science Museum/Science and Society Picture
Library, (bottom) Paul Schemt, Eye Ubiquitous

CONTENTS

BEFORE NUMBERS

Numbers and counting are such an important part of our lives that it is hard to imagine a world without them. But for thousands of years, there was no written system for recording numbers. Instead, people used a mixture of experience and guesswork to estimate and compare quantities and sizes, and they invented simple ways of keeping count of their livestock, sacks of grain, and other belongings. Written numbers developed from these unsophisticated ways of keeping tally, which varied from one part of the world to another.

HEAD COUNT

Before written numbers were invented, people had various ways of estimating amounts. For example, two similar-sized herds of cattle might be driven through a gate, a pair at a time, to figure out which herd was bigger. But people could only "count" objects they could see. They had no way of dealing with imaginary calculations.

In many parts of Africa, cattle are important status symbols. The size of a herd determines a person's standing in the community. Cattle are also used instead of money for bartering and exchange.

Tally sticks

KEEPING A TALLY

One of the first ways of recording amounts was to keep a tally. Each item was marked with a notch carved on a stick, stone, bone, or piece of clay. Tallies could also be kept by adding another pebble, shell, or bead to a pile. Tallies were the most common way of keeping count for thousands of years.

BODY COUNT

Another way of keeping count was to point to different parts of the body. For thousands of years, people around the world kept count on their 10 fingers. Today, our counting system is still based on the number 10, and we still learn to count on our hands!

A tribe from Papua New Guinea took this way of counting even further. They began counting from the little finger on the right hand (this counted as one) through the other fingers; to the right wrist, elbow, and shoulder; to the right ear and right eye; then to the left eye, left ear, and down to the little finger on the left hand (making 20).

TIED UP IN KNOTS

The Incas of Peru recorded numbers of people, animals, and payment of taxes by tying knots in lengths of colored string, called quipus. The number and position of the knots indicated different amounts.

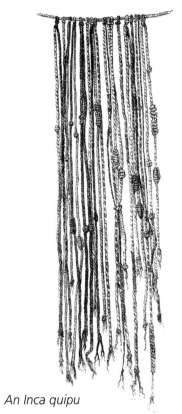

An Inca quipu

SIGNPOST

Numbers have a huge impact on everything we do, from telling the time, making phone calls, and counting money to choosing the right size of clothes and shoes to wear. Calculations involving numbers form the very basis of mathematics, engineering, astronomy, and accounting. You might be surprised at how often you use numbers in your everyday life. Even when you're waiting for a bus or train, how do you know which is the right one to catch?

Millions of calculations are made in a day in stock exchanges around the world.

IN FACT...

The Nambiquara people who live on the edge of the Amazon rain forest in Brazil do not have any system of numbers or counting. Instead, they have a verb that describes similar amounts.

THE FIRST NUMBERS

It took thousands of years to develop written symbols to stand for numbers and to work out rules of counting. The earliest number systems were devised about 5,000 years ago (shortly after writing began) in the ancient Middle East area. At this time, the first cities grew up and people began to trade far and wide. Traders needed written numbers to keep more accurate accounts, keep track of goods, and calculate taxes that were paid to the ruling king.

BREAKTHROUGH

One of the very first written number systems was invented by the Sumerians (who lived in the area now known as Iraq) in about 3000 B.C. They recorded numbers as differently shaped marks, pressed into small clay tablets. The Sumerians were also the first people to use writing.

NUMBERS AND NUMERALS

Symbols that stand for numbers are called numerals. A number system is a set of numerals and rules for use in counting, calculating, comparing different amounts, measuring things, and sending messages. Different peoples devised different number systems at different times. The Sumerian number system was very simple. It had only two numerals, for 1 and 10. They used two number systems: one based on counting in 60s, the other on counting in 10s.

BABYLONIAN BILLS

The Babylonians took over the Sumerian kingdom, with its number and writing systems, 4,000 years ago. Hundreds of baked clay tablets survive from that time, showing bills, receipts, and contracts. Archaeologists have even found tables that were used to solve multiplication problems.

Very early Sumerian writing, including the marks for numbers, on a clay tablet. Tablets like this show us that the people who wrote them were clever mathematicians.

EGYPTIAN ADDITION

The ancient Egyptians used picture symbols, called hieroglyphs, to write words and numbers. The symbol for 1 was a papyrus leaf; for 10, a bent-over papyrus leaf; for 100, a coiled rope; for 1,000, a lotus flower; for 10,000, a snake; for 100,000, a tadpole; and for 1 million, a scribe. To write the number 30, for example, you drew three bent-over papyrus leaves (3 x 10). Numbers were written from right to left.

COUNTING LESSONS

Ancient Egyptian schoolchildren learned to count and do sums through games. Their teachers used these games to show them how to use numbers for practical purposes, such as organizing soldiers and supplies in an army, surveying agricultural land, or running a household.

Lotus flowers were sacred in ancient Egypt. The Egyptians believed that the world was created by a god who appeared from a lotus. The flower shape was used to represent the number 1,000.

IN FACT...

Our main source of information about Egyptian numbers is a papyrus document called the Rhind Papyrus (named after A. H. Rhind, who bought it in Egypt in the 1850s). It was written in about 1575 B.C. by a scribe named Ahmes, who copied it from an even older papyrus document. Among other things, the papyrus describes how to use fractions to divide the daily ration of bread and beer among temple workers. It is one of the oldest mathematical records in the world.

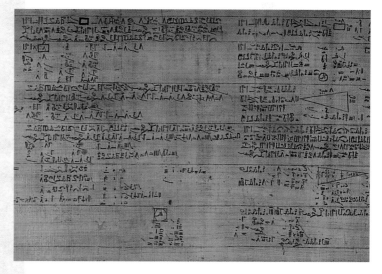

Part of the Rhind Papyrus showing calculations about the areas of triangles

GREEK NUMBERS

The ancient Greeks had two different number systems. The first one was a simple system based on units of 5 and 10. Each number was represented by the first letter of the Greek word for that number. By the 1st century B.C., this had been replaced by a more complicated number system, where each of the 27 letters of the Greek alphabet was used as a number.

Roman numerals on a clock face

ROMAN NUMERALS

The number system used by the Romans was based on seven signs: I for 1; V for 5; X for 10; L for 50; C for 100; D for 500; and M for 1,000. Each numeral was made up of straight lines, so that they were easy to carve into stone or wax. Roman numerals were devised over 2,000 years ago. They are still sometimes used today for dates and for numbers on clock faces.

*Part of an Aztec manuscript showing the dots and dashes used to write numbers. The Aztecs used a similar system to the Maya. (See the **In Fact** box on page 11.)*

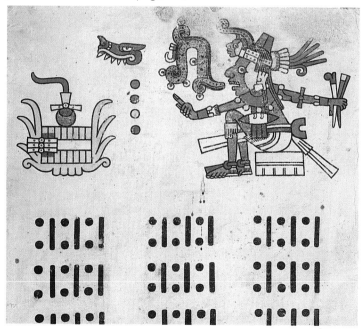

SIGNPOST

Roman numerals are combined to make larger or smaller numbers. If the numeral on the right is smaller than, or equal to, the one on the left, you add them. So V (5) followed by I (1) would equal 6. If the numeral on the left is smaller, you subtract it. So I (1) followed by V (5) would equal 4. Try working out these Roman totals:

MM   XXXVII
XL   LIV
MCMXCVI

Answers: MM = 2000, XXXVII = 37, XL = 40, LIV = 54, MCMXCVI = 1996

Playing cards – which use numbers – were invented in China by the 9th century A.D.

CHINESE SYSTEMS

The ancient Chinese had five sets of numerals, each used for a different purpose. The most important were stick numerals, based on the wooden sticks used on a counting board; basic numerals; and "official" numerals. Official numerals were used on banknotes and official documents. They were very elaborate, to make forgery more difficult.

DIFFERENT NUMERALS

Different peoples invented their own number systems and their own numerals. Here is a selection from around the world:

Babylonian	𒐗	𒐚	𒐜	𒐝	𒐞
Egyptian	ꓲꓲꓲꓲ	ꓲꓲ꓾ꓲꓲꓲ	ꓲꓲꓲ	ꓲꓲꓲꓲ	ꓲꓲꓲꓲꓲ
Greek	Δ	E	F	Z	H
Roman	IV	V	VI	VII	VIII
Ancient Chinese	四	五	六	七	八
Mayan	••••	▬	⩻	⩻⩻	⩻⩻
Hindu	୪	୫	୬	୭	୮
Arabic	୪	୫	୬	୭	୮
Modern (based on Arabic numerals)	4	5	6	7	8

┃N FACT...

In the 4th century B.C. the Maya of Central America invented their own system of picture writing. Numbers were written as a series of dots and dashes. The Maya were excellent mathematicians. They used their knowledge of numbers to devise an accurate 365-day calendar, long before other people. The Maya were also among the first to have a symbol for zero.

NUMBERS MOVE ON

The numerals we use today are based on those invented in India by Hindu mathematicians in about 500 A.D. They are often called "Arabic" numerals because Arab traders and scholars learned them from the Hindus, then spread their use to the West in about 1200 A.D. Until then, most people in Europe used Roman numerals. The Arabic system was much quicker and easier to use.

A painting of an early Arab trading ship. Arab traders helped to spread their number system to the West.

CHANGING NUMBERS

Arabic numerals were slow to catch on in Europe. Most people could not read or write so could not use them. Mathematicians using Roman numbers were respected for their skill, and many feared that an easier number system would rob them of their high standing. In 1299 a decree was issued in Italy forbidding bankers to use the new numerals. It was only in the 17th century, when merchants began to use Arabic numerals for their accounts, that they became more popular and widespread.

BACK TO BASE

We count in multiples, or groups, of 10, probably because we have 10 fingers on our hands (see page 7). This is called base 10, or the decimal system. But the Sumerians and Babylonians used base 60, which we still use today for measuring time in minutes and seconds and in some branches of mathematics.

Counting in 10s

PLACE VALUE

The Hindu-Arabic number system made calculations easier because the value of a number depended on its position as well as on the number itself. This is called place value. In the Roman system, numbers had the same value wherever they appeared, which could be very confusing. Mathematicians using the new system found that they could experiment with far more complicated math problems than ever before because the Arabic system was so much easier to use.

TESTING 1, 2

Computers use base 2, or the binary system, in their calculations. In binary, any number can be shown using two symbols, 0 and 1. This makes numbers simpler but longer than decimal numbers. For example, 9 becomes 1001; 10 becomes 1010. Computers can use the binary system because they have electronic systems that give them big enough memories to deal with extremely large numbers.

Mathematicians, astronomers, and scholars found the Arabic number system easier to use.

SPECIAL NUMBERS

Some numbers are considered particularly special or important. This may have to do with mathematics, magic, or superstition. Some numbers are given special names to indicate their value or mathematical properties. Some numbers are thought to be sacred; others are lucky or unlucky. People often have their own favorite numbers that are lucky for them.

The Russian novelist Leo Tolstoy (on left, playing chess) thought the number 28 was his lucky number. He was born on August 28, 1828.

LUCKY 13

Many people think that 13 is an unlucky number. To the Maya, however, it was one of the most sacred numbers of all because it represented the 13 original Mayan gods. The Maya also believed that the world and sky were arranged in 13 layers.

GREEK BELIEFS

The ancient Greeks were very superstitious about numbers. They believed that all numbers were sacred but that some were friendly and others evil. They saw the number 8 as the symbol of death while 10 was the number of harmony. Odd numbers were thought to be female and even numbers to be male.

This page of a Mayan manuscript shows special dates (a dot means 1, a stroke means 5) and pictures of a great god.

NAMES OF NUMBERS

In mathematics, numbers are given different names, depending on their special properties. Here are a few of these names:

Real numbers – the numbers we use for counting

Prime numbers – numbers that can be divided only by themselves and by one: for example, 3, 5, 11

Negative numbers – numbers below zero

Perfect numbers – numbers that equal the total of all the numbers that can be divided into them. For example, 28 can be divided by 1, 2, 4, 7, and 14. It is a perfect number because 1 + 2 + 4 + 7 + 14 = 28.

BOOK OF NUMBERS

The fourth book of the Bible is called the Book of Numbers. It tells how Moses was commanded by God to count the Israelites to discover such details as the number of fighting men over 20 years old, and so on. This type of numerical survey is called a census.

Moses leaving Mount Sinai after being told by God to count the Israelites

SIGNPOST

A magic square is an arrangement of numbers in which every row, column, and diagonal adds up to the same total. One of the oldest is said to have been revealed to a Chinese emperor in 2000 B.C. It was marked on the back of a magical tortoise and was interpreted as a message from the gods telling the emperor how to establish order in his empire.

IN FACT...

A million is 1 followed by six zeros. A trillion is a million million (1 followed by 12 zeros). A centillion is 1 followed by 303 zeros. A googol is 1 followed by 100 zeros. This name was first used by an American mathematician, Edward Kasner, who ran out of names ending in "-illion." He asked his nine-year-old nephew, who came up with the word *googol*!

WORKING NUMBERS

One of the most important uses of numbers is in mathematics. The Sumerians, Babylonians, and ancient Egyptians were the first mathematicians and highly skilled scientists. They used mathematics to solve practical problems of trade, telling the time, farming, building, and astronomy.

BASIC ARITHMETIC

The basic calculating techniques of addition, subtraction, multiplication, and division are known as arithmetic, from the Greek word *arithmos,* which means "number."

BRANCHES OF MATH

There are many different branches of math. Here are some of the main ones:

Algebra – the study of number systems and the properties of numbers. It often uses letters or symbols instead of numbers.

Calculus – deals with problems about incredibly small changes in areas, distances, or times. It was devised independently by two mathematicians in the 17th century: Isaac Newton in England and Gottfried Wilhelm von Leibniz in Germany.

Geometry – the study of an object's shape and space.

Trigonometry – the study of triangles.

To the Greeks, arithmetic was a very special subject and only educated, free Greek citizens could use it. It was forbidden to slaves.

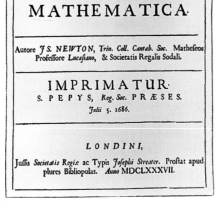

PHILOSOPHIÆ

NATURALIS

PRINCIPIA

MATHEMATICA·

Autore *JS. NEWTON,* Trin. Coll. Cantab. Soc. Matheseos Professore *Lucasiano,* & Societatis Regalis Sodali.

IMPRIMATUR·
S. PEPYS, *Reg. Soc.* PRÆSES.
Julii 5. 1686.

LONDINI,

Jussu *Societatis Regiæ* ac Typis *Josephi Streater.* Prostat apud plures Bibliopolas. *Anno* MDCLXXXVII.

SIGNS AND SYMBOLS

Apart from numerals themselves, many different signs and symbols are used in mathematics as a shorthand way of showing how numbers should be used. The most basic are: + (plus), - (minus), x (multiplied by), ÷ (divided by), and = (equals). The first known use of the = sign was in England in 1557.

The title page of one of Isaac Newton's great works on mathematics

The word *geometry* comes from the Greek for "earth measuring." In about 240 B.C., a Greek astronomer, Eratosthenes, used geometry to calculate the size of the earth. He measured in stadia. (A stadium is an ancient unit of measurement.) Eratosthenes' calculations show him to have been amazingly near modern measurements of the earth's circumference around the poles (24,800 miles or 40,007 kilometers).

VALUES OF PI

The number pi (π) is used to express the ratio between the circumference of any circle and its diameter. For most purposes, pi is said to equal 3.14. Computers have calculated pi to 1 billion decimal places (this means there are 1 billion numbers after the decimal point).

The Greek mathematician Archimedes (c. 287-212 B.C.) was the first to calculate the value of pi.

SIGNPOST

The ancient Egyptians used their knowledge of geometry to help them design the pyramids. They were remarkably accurate. The sides of the Great Pyramid, at Giza, are of equal length to within 1 inch (3 centimeters), and the angles at the corners are equal to within a fraction of a degree.

MATHEMATICAL MINDS

The ancient Greeks produced many great mathematicians whose discoveries are still in use today. In other parts of the world, Chinese, Hindu, and Arab scholars developed mathematical theories that influenced many European mathematicians. Europeans such as René Descartes and Isaac Newton were leaders in mathematics in the 17th century. The period of modern mathematics began in the 19th century. Mathematicians continue their work today, trying to unravel the mysteries of numbers.

EUCLID'S ELEMENTS

In about 330 B.C., the Greek mathematician Euclid (left) published his ideas on geometry and number theory in a book called *Elements*. In it he collected together the work of other Greek mathematicians, such as Thales and Pythagoras, and for the first time set out 10 of the most basic mathematical rules. Geometry students used *Elements* as their textbook for more than 2,000 years, and it still influences the way mathematicians think today.

THE PYTHAGOREAN THEOREM

The Greek philosopher and mathematician Pythagoras (c. 560-480 B.C.) is famous for the theorem that takes his name. He found a way of calculating the lengths of sides of right-angled triangles. Using his rule, once you know the length of two sides, you can figure out the length of the third.

The Pythagorean theorem states that, in a right-angled triangle, the square of the longest side (the hypotenuse) equals the square of the other two sides added together.

HYPATIA OF ALEXANDRIA

Hypatia (370-415) was one of the very few women scientists of the ancient world. She taught mathematics and philosophy at the famous Museum (university) in Alexandria, Egypt. According to legend, she was so beautiful that she had to lecture from behind a screen so that her pupils would not be distracted from their studies.

A page of calculations by the 19th-century French mathematician Evariste Galois

DEDICATION

The French mathematician Evariste Galois (1811-32) was killed in a duel in the early 19th century. He was so dedicated to his research that he spent the night before the duel scribbling down details of his latest theory. The Galois theory has inspired many modern theories about algebra.

ECCENTRIC GENIUS

The English mathematician Alan Turing (1912-54) was brilliant, and eccentric. He taught himself to knit gloves for winter. But he never learned how to finish off the ends of the fingers, and so his gloves always had long strands of wool dangling from them! (For more about Turing, see page 27.)

Alan Turing

BREAKTHROUGH

Isaac Newton's great work, *Principia Mathematica*, was published in 1687 (see page 16). Using mathematical calculations, he showed that the force of gravity pulls objects toward the earth and dictates how the planets orbit the sun. The story goes that he realized this when he saw an apple fall from a tree. Newton (1642-1727) is considered one of the greatest scientists of all time.

IN FACT...

In the 18th century, mathematicians were often connected to a university or royal court. The Swiss mathematician Leonhard Euler (1707-83) was employed by Frederick the Great of Prussia. Euler produced so much work that his papers were still being published for the first time 50 years after his death.

BEFORE MONEY

The history of numbers and counting is closely linked to the history of money. In ancient times, before numbers and money were invented, people bartered (exchanged) one set of goods for another. As cities and trade grew, however, they needed better, more fixed ways of buying and selling and of keeping their accounts. This is where money and counting systems came into their own.

Feather money from the Pacific Islands

EXCHANGING GOODS

Before people had coins and paper money, they paid for goods with other valuable objects. These included shells, beads, spears, animal teeth, lumps of metal, gold and silver rings, feathers, and stones. The Aztecs of Mexico used cocoa beans as currency.

A cowrie shell

SHELL MONEY

For centuries cowrie shells were used as money in Africa and Asia. In North America, Indians wove tiny clam and dentalia shells into necklaces, belts, and headdresses that could be exchanged for other goods.

RODS AND RINGS

In Africa metal rods and rings were a common form of money until the 1940s. The Ibo people of Nigeria used copper rings, called *manillas*, to make payments. In Central Africa 2 U-shaped copper rods, called *kongas*, could buy you a male slave, 3 a female slave, and 10 a wife!

The Yap islanders of the Pacific used huge stone disks as money. The largest were 13 feet (4 meters) wide!

Camels can be a sign of wealth and used as "money."

COUNTING CAMELS

In some countries animals such as pigs, cattle, and camels are still used as currency. In the Middle East, camels are a sign of wealth and status. They are traded for other goods, given as gifts, and used to pay fines and as part of a bride's wedding dowry.

HOE COINS

In ancient China, people used bronze miniatures of tools to pay for the real thing. Miniature hoe "coins" bought real hoes, miniature spade or knife "coins" bought real spades and knives, and so on.

An ancient Chinese bronze "coin" in the form of a blade

IN FACT...

Until just over 100 years ago, tea was used as money in Tibet and China. Tea and wood shavings were pressed into a brick shape worth a certain amount of money. Small pieces could be broken off as change.

MAKING MONEY

Lumps of metal, such as gold, silver, and copper, were being used as money more than 4,000 years ago. The value of the metal lump depended on its weight, not on its size or shape. By the 10th century B.C., these lumps of metal had developed into coins. Buying and selling with coins was much easier than bartering. You simply had to count out the correct number of coins for something rather than find another equally valuable object to exchange.

Lydian coins from the 7th century B.C.

FIRST COINS

The first coins were minted in Lydia (Turkey) in the 7th century B.C. They were lumps of electrum (a mixture of silver and gold) stamped with the lion's head seal of the king. They were originally used for paying state officials but soon became invaluable for trade.

PAPER MONEY

The first paper money was invented in China. By the 11th century, handwritten notes were in regular use, replacing the heavy iron coins of the time. The first European banknotes were issued in Sweden in 1661 because there was a shortage of silver coins.

The Mongol emperor Kublai Khan supervises the distribution of early Chinese paper money.

COIN CLIPPING

In the Middle Ages, coins were made by hand and were never perfectly round. To make a bit of extra money, people clipped the edges off coins and melted them down to make new coins. The punishment for coin clipping was to have your right hand cut off. Later, coins were given "milled" edges, where grooves or inscriptions were etched on to the edge of the coin. This helped to stop people from clipping them. Today, milled edges are used to prevent forgery and to help blind people recognize different coins.

A 14th-century gold coin with parts of its edges clipped off

Italian bankers in the 14th century

BANK BALANCE

The first modern banks were set up in Italy in the Middle Ages. The term *bank* comes from the Italian word *banco*, which means "bench." This refers to the benches in the marketplace on which the moneylenders sat.

SIGNPOST

Many of the world's currencies, such as the Spanish peseta, British pound, and Italian lira, are named after units of weight. These were used to weigh the original lumps of metal that were the ancestors of coins.

EARLY CALCULATORS

When people began counting with numbers, they did not have adding machines or calculators to help them. They worked things out in their heads, with sticks or stones, or on their fingers and toes. Finger calculating was used by people all over the world for thousands of years and is still used today. The earliest "calculators" were not machines but people whose job it was to carry out calculations.

BEADS AND BARS

One of the earliest calculating devices was the abacus. Bead-framed abacuses have been used for thousands of years in China, Japan, and Russia. They consist of a wooden frame with beads strung on wires or bars. The beads represent units, tens, hundreds, and so on. You slide the beads up and down to show different amounts.

A Chinese abacus

COUNTER CASTING

Counter casting was a way of doing calculations too complicated to work out in your head. The counters were laid out and moved around on a board, marked to indicate different place values. The Greeks and Romans used *calculi*, or pebbles, as counters. Medieval accountants used brass counters called jettons. Counter casting was used for thousands of years before being replaced by "pen-reckoning" (writing out calculations on paper).

From abacus to calculator!

Some people can perform extraordinary feats of mental arithmetic. An Indian woman, Shakuntala Devi, multiplied 7,686,369,774,870 by 2,465,099,745,779 in her head, in just 28 seconds! She gave the correct answer of 18,947,668,177,995,426,462,773,730.

Until the end of the 18th century, the English Treasury used counter casting for all its royal accounts and taxes. The counters were laid on a large table, called the Exchequer Table because it looked like a giant chess board.

LOG TABLES

Logarithms are special numbers that allow you to do complicated multiplication and division simply by adding two numbers together. You can look up the logarithms for any numbers in books of tables. The first such tables were published in 1614 by the Scottish mathematician John Napier (1550 -1617).

Logarithms were used well into the 1970s, when they were largely replaced by electronic calculators.

John Napier

The Exchequer in action in the 15th century

Mechanical Means

All calculating devices, from abacuses to logarithms to computers, are designed to make our use of numbers quicker and easier. The first mechanical calculating machines appeared in the 17th century. Today, we have a wide range of sophisticated calculating technology, from electronic calculators to powerful computers. These have increased speed and accuracy even further.

SLIDING RULES

An early slide rule from the mid-15th century

Until the invention of pocket calculators, slide rules were the most useful calculating devices. They use logarithms and a sliding scale to give answers. Most slide rules today are made of plastic. When they first appeared in the early 17th century, they were made of wood, ivory, and brass.

THE ANALYTICAL ENGINE

In 1834 the English scientist Charles Babbage (1792-1871) designed a mechanical calculator called the Analytical Engine. Not only could it perform one calculation a second, it also had a memory for storing answers. Countess Ada Lovelace (sister of the poet Byron) wrote programs that were fed into the machine on a series of punched cards. The Analytical Engine was the forerunner of modern computers, but it was too expensive to build and was never finished.

COMPUTER COUNTING

The first fully electronic computing machine was ENIAC (Electronic Numerical Integrator and Calculator). It was built at the University of Pennsylvania in 1946. ENIAC was a truly gigantic calculator. It weighed about 30 tons and could perform 5,000 calculations a second. The latest computers can perform hundreds of millions of calculations a second. Today, the same work can be done by a calculator small enough to fit into your pocket!

Charles Babbage, the inventor of the Analytical Engine and one of the greatest mathematicians of his day

In World War II the British intelligence service used the calculating power of a computer called COLOSSUS to crack enemy codes. COLOSSUS could read and decipher thousands of signals a second. It was invented and operated by the mathematical genius Alan Turing (see page 19).

POCKET CALCULATORS

The first electronic pocket calculators were produced in the United States in 1972. All the electronic equipment needed to perform a wide range of calculations is packed onto a tiny silicon chip.

In 1642 the French mathematician Blaise Pascal (1623-62) invented the first effective mechanical adding machine. It could add and subtract automatically, using a series of numbered cogwheels, connected by gears and chains.

A replica of Pascal's calculating machine from 1642

Modern calculators are small and easy to carry and use, but many people around the world still use the traditional abacus.

TIMELINE

B.C.
— **c. 3000** First number system devised in Sumeria
— **c. 3000** Abacus invented in China
— **c. 2000** Babylonians use multiplication tables
— **c. 1575** The Rhind Papyrus is written by the Egyptian scribe Ahmes
— **c. 700** First coins are minted in Lydia (Turkey)
— **c. 560-480** Life of Pythagoras
— **330** Euclid publishes *Elements*

A.D.
— **370-415** Life of Hypatia of Alexandria
— **c. 500** Hindu mathematicians invent a new set of numerals and a symbol for zero
— **c. 500s** Paper money first issued in China
— **c. 1200** Arabs bring Hindu numbers to Europe
— **1500s** Decimal point first used in Europe
— **1557** Equals sign (=) first used
— **1614** First logarithm tables published by John Napier
— **1622** Slide rule invented by English clergyman William Oughtred
— **1642** Blaise Pascal invents the first mechanical adding machine
— **1661** First European banknote issued in Sweden
— **1687** Isaac Newton publishes *Principia Mathematica*
— **1834** Charles Babbage designs the Analytical Engine
— **1945** ENIAC is built in the United States
— **1972** First electronic pocket calculators produced

GLOSSARY

Arithmetic The basic processes of calculation: addition, subtraction, division, and multiplication.

Binary system A counting system based on groups of two.

Census A survey of the numbers of people in a particular country or town, or in a particular job and so on.

Circumference The earth's circumference is the distance around the earth. A circle's circumference is the distance around a circle.

Currency The type of money in current use in a country.

Decimal system A counting system based on groups of 10.

Diameter The distance across a circle, measured through the center.

Googol A very large number, written as 1 followed by 100 zeros.

Gravity The force that attracts two objects toward each other.

Hieroglyphs Picture symbols used a form of writing by the ancient Egyptians.

Hypotenuse The longest side of a right-angled triangle.

Mental arithmetic Calculations done in a person's head, without the help of any calculators or counting aids.

Minted How a coin is made by stamping it out of a sheet of metal.

Numeral A written symbol that stands for a particular number.

Papyrus A reed grown in ancient Egypt. Used as a writing material before the invention of paper.

Pi (π) The ratio between the circumference of a circle and its diameter, or 3.14.

Quipu A counting device consisting of lengths of colored string, used by the Incas of Peru.

Ratio The mathematical relationship between two amounts, based on the number of times one divides into the other.

Square The square of a number is the number multiplied by itself. For example, the square of 4 is 16 (4 x 4).

Surveying Inspecting and measuring a piece of land to see if it is big enough and suitable for building on.

Tally A piece of wood, bone, or clay marked with notches to keep count of a number of objects.

Theorem A mathematical rule expressed by symbols in an equation.

INDEX